Contents

I0503315

Executive Summary i

I.   Introduction 1

II. Background 3

    A. Face Facts Workshop 3

       1. Recent advances in facial recognition technologies 3

       2. Current commercial uses of facial recognition technologies 4

       3. Possible future uses of facial recognition technologies 6

       4. Privacy concerns raised by current and possible future uses of
          facial recognition technologies 7

    B. Public Comments 8

III. General Themes 9

IV. Case Studies on Common Commercial Uses of Facial
    Recognition Technologies 11

      Case Study #1: Facial Detection 11

      Case Study #2: Detection or Recognition of Demographic
        Characteristics in Digital Signs 13

      Case Study #3: Facial Recognition in Online Social Networks 17

V.   Conclusion 21

Dissenting Statement of Commissioner J. Thomas Rosch A1

Executive Summary

> "John Anderton... You could use a Guinness right about now."
> – Scene from the film Minority Report depicting use of
> biometric technology to target individualized ads[1]

In the 2002 film Minority Report, Steven Spielberg imagined a world in which companies use biometric technology to identify us and serve us targeted ads. Ten years later, that vision is coming closer to reality. Having overcome the high costs and poor accuracy that once stunted its growth, one form of biometric technology – facial recognition – is quickly moving out of the realm of science fiction and into the commercial marketplace.

Today, companies are deploying facial recognition technologies in a wide array of contexts, reflecting a spectrum of increasing technological sophistication. At the simplest level, the technology can be used for facial detection; that is, merely to detect and locate a face in a photo. Current uses of facial detection include refining search engine results to include only those results that contain a face; locating faces in images in order to blur them; ensuring that the frame for a video chat feed actually includes a face; or developing virtual eyeglass fitting systems and virtual makeover tools that allow consumers to upload their photos online and "try on" a pair of glasses or a new hairstyle.

A more refined version of facial recognition technology allows companies to assess characteristics of facial images. For instance, companies can identify moods or emotions from facial expressions to determine a player's engagement with a video game or a viewer's excitement during a movie. Companies can also place cameras into digital signs to determine the demographic characteristics of a face – such as age range and gender – and deliver targeted advertisements in real-time in retail spaces.

In the most advanced application, companies can use the technology to compare individuals' facial characteristics across different images in order to identify them. In this application, an image of an individual is matched with another image of the same individual. If the face in either of the two images is identified – that is, the name of the individual is known – then, in addition to being able to demonstrate a match between two faces, the technology can be used to identify previously anonymous faces. This is the use of facial recognition that potentially

---

1. *Minority Report*, Dir. Steven Spielberg, DreamWorks, 20th Century Fox, 2002.

raises the most serious privacy concerns because it can identify anonymous individuals in images. One prevalent current use of this application is to enable semi-automated photo tagging or photo organization on social networks and in photo management applications. On social networks this feature typically works by scanning new photos a user uploads against existing "tagged" photos. The social network then identifies the user's "friends" in the new photos so the user can tag them.

On December 8, 2011, the Federal Trade Commission ("FTC" or "Commission") hosted a workshop – "Face Facts: A Forum on Facial Recognition Technology" ("Face Facts workshop") – to explore developments in this rapidly evolving field. Panelists discussed a number of issues, including: recent advances in facial recognition technologies; current and possible future commercial uses of facial recognition technologies; ways consumers can benefit from these uses; and privacy and security concerns raised. Following the workshop, the FTC received eighty public comments discussing these issues from private citizens, industry representatives, trade groups, consumer and privacy advocates, think tanks, and members of Congress. In this report, FTC staff has synthesized those discussions and comments in order to develop recommended best practices for protecting consumer privacy in this area, while promoting innovation.

To begin, staff recommends that companies using facial recognition technologies design their services with privacy in mind, that is, by implementing "privacy by design," in a number of ways. First, companies should maintain reasonable data security protections for consumers' images and the biometric information collected from those images to enable facial recognition (for example, unique measurements such as size of features or distance between the eyes or the ears). As the increasing public availability of identified images online has been a major factor in the increasing commercial viability of facial recognition technologies, companies that store such images should consider putting protections in place that would prevent unauthorized scraping which can lead to unintended secondary uses. Second, companies should establish and maintain appropriate retention and disposal practices for the consumer images and biometric data that they collect. For example, if a consumer creates an account on a website that allows her to virtually "try on" eyeglasses, uploads photos to that website, and then later deletes her account on the website, the photos are no longer necessary and should be discarded. Third, companies should consider the sensitivity of information when developing their facial recognition products and services. For instance, companies developing digital signs equipped with cameras using facial recognition technologies should consider carefully where to place such signs and avoid placing them in sensitive areas, such as bathrooms, locker rooms, health care facilities, or places where children congregate.

Staff also recommends several ways for companies using facial recognition technologies to provide consumers with simplified choices and increase the transparency of their practices. For example, companies using digital signs capable of demographic detection – which often look no different than digital signs that do not contain cameras – should provide clear notice to consumers that the technologies are in use, before consumers come into contact with the signs. Similarly, social networks using a facial recognition feature should provide users with a clear notice – outside of a privacy policy – about how the feature works, what data it collects, and how it will use the data. Social networks should also provide consumers with (1) an easy to find, meaningful choice not to have their biometric data collected and used for facial recognition; and (2) the ability to turn off the feature at any time and delete any biometric data previously collected from their tagged photos.

Finally, there are at least two scenarios in which companies should obtain consumers' affirmative express consent before collecting or using biometric data from facial images. First, they should obtain a consumer's affirmative express consent before using a consumer's image or any biometric data derived from that image in a materially different manner than they represented when they collected the data. Second, companies should not use facial recognition to identify anonymous images of a consumer to someone who could not otherwise identify him or her, without obtaining the consumer's affirmative express consent. Consider the example of a mobile app that allows users to identify strangers in public places, such as on the street or in a bar. If such an app were to exist, a stranger could surreptitiously use the camera on his mobile phone to take a photo of an individual who is walking to work or meeting a friend for a drink and learn that individual's identity – and possibly more information, such as her address – without the individual even being aware that her photo was taken. Given the significant privacy and safety risks that such an app would raise, only consumers who have affirmatively chosen to participate in such a system should be identified.

The recommended best practices contained in this report are intended to provide guidance to commercial entities that are using or plan to use facial recognition technologies in their products and services. However, to the extent the recommended best practices go beyond existing legal requirements, they are not intended to serve as a template for law enforcement actions or regulations under laws currently enforced by the FTC. If companies consider the issues of privacy by design, meaningful choice, and transparency at this early stage, it will help ensure that this industry develops in a way that encourages companies to offer innovative new benefits to consumers and respect their privacy interests.

I. Introduction

As facial recognition technologies have become more accurate and less costly, commercial interest and investment in these technologies has grown.[2] The ability to make inferences about an individual based on his or her unique mix of facial characteristics can have countless uses, many of which are innovative and beneficial to consumers. However, the rapidly expanding commercial use of these technologies, particularly when combined with the growing availability of identified images online, can also pose complex privacy issues. Recognizing that the commercial use of these technologies will likely continue to grow, the FTC has sought to understand how they are being used, how they could be used, and the potential risks and benefits of such technologies.

The FTC's December 2011 Face Facts workshop was a first step towards exploring new developments in this field and their potential impact on consumers.[3] The workshop featured panel discussions on the capabilities of facial recognition technologies, current and potential implementations of these technologies, and the benefits and privacy concerns these uses can generate. The workshop was followed by a one month public comment period in which Commission staff sought further input and insight on these issues.

This report builds upon the discussions at the Face Facts workshop and the written comments received thereafter to set forth a series of case studies illustrating recommended best practices for companies using or planning to use facial recognition technologies in their products or services.[4] These best practices draw upon the three core principles outlined in the FTC's March 2012 report, "Protecting Consumer Privacy in an Era of Rapid Change" ("Privacy Report").[5]

---

2. Throughout this report, staff uses the term "facial recognition" to broadly refer to any technology that is used to extract data from facial images. *See* Sony, Face Recognition Technology, http://www.sony.net/SonyInfo/technology/technology/theme/sface_01.html.

3. FTC Workshop, *Face Facts: A Forum on Facial Recognition Technology* (Dec. 8, 2011), http://www.ftc.gov/bcp/workshops/facefacts/. The Commission recognizes that there are many forms of biometric technology – fingerprints, retinal scans, voice-prints, etc. – that raise similar issues as facial recognition technology. However, the workshop and this report focus solely on facial recognition.

4. This report addresses solely commercial uses and does not address the use of facial recognition technologies for security purposes or by law enforcement or government actors.

5. FTC, Protecting Consumer Privacy in an Era of Rapid Change, Recommendations for Businesses and Policymakers, FTC Report (Mar. 2012), *available at* http://www.ftc.gov/os/2012/03/120326privacyreport.pdf. Commissioner Rosch dissented from the issuance of the Final Privacy Report. *See id.* at Appendix C.

These principles are:

1. **Privacy by Design:** Companies should build in privacy at every stage of product development.

2. **Simplified Consumer Choice:** For practices that are not consistent with the context of a transaction or a consumer's relationship with a business, companies should provide consumers with choices at a relevant time and context.

3. **Transparency:** Companies should make information collection and use practices transparent.

This report begins by providing background information from the Face Facts workshop and discussing the public comments. Next, it addresses general themes that panelists and commenters raised. Finally, it explores a series of case studies, each focused on a common commercial use of facial recognition technologies. The recommended best practices demonstrated in the case studies are intended to provide guidance to commercial entities that are using or plan to use facial recognition technologies in their products and services. However, to the extent the recommended best practices go beyond existing legal requirements, they are not intended to serve as a template for law enforcement actions or regulations under laws currently enforced by the FTC.[6]

---

6. Under Section 5 of the FTC Act, the Commission is authorized to take action against unfair or deceptive acts or practices. 15 U.S.C. § 45(a). Unfair acts or practices are defined as those that cause or are likely to cause substantial injury to consumers which is not reasonably avoidable by consumers themselves and not outweighed by countervailing benefits to consumers or to competition. 15 U.S.C. § 45(n). If a company uses facial recognition technologies in a manner that is unfair under this definition, or that constitutes a deceptive act or practice, the Commission can bring an enforcement action under Section 5. In contrast, in other countries and jurisdictions, such as the European Union, in certain circumstances, consumers may need to be notified and give their consent before a company can legally use facial recognition technologies. *See Face Facts Workshop, Remarks of Simon Rice, Technology Information Commissioner's Office, United Kingdom,* at 186, 193.

II. Background

A. Face Facts Workshop

Researchers, academics, industry representatives, and consumer and privacy professionals all took part in a series of wide-ranging discussions at the Face Facts workshop. The facial recognition technologies discussed included technologies that merely detect basic human facial geometry; technologies that analyze facial geometry to predict demographic characteristics, expression, or emotions; and technologies that measure unique facial biometrics.[7] Major topics included: (1) recent advances in facial recognition technologies, (2) current commercial uses of facial recognition technologies, (3) possible future uses of facial recognition technologies, and (4) privacy concerns raised by current and possible future uses of facial recognition technologies.

1. Recent advances in facial recognition technologies

Until recently, because of high costs and limited accuracy, companies have not used facial recognition technologies on a widespread basis. However, recent years have brought steady improvements in these technologies. For example, from 1993 to 2010, tests conducted by the National Institute of Standards and Technology ("NIST") showed that the false reject rate – the rate at which facial recognition systems incorrectly rejected a match between two faces that are, in fact, the same – was reduced by half every two years.[8] In 2010, in controlled tests, the error rate stood at less than one percent.[9]

Workshop panelists identified several developments that have contributed to the increased accuracy in facial recognition systems. For example, better quality digital cameras and lenses create higher quality images, from which biometric data can be more easily extracted.[10] In addition, the goal of some facial recognition technologies is to match an image of an unknown

---

7. The biometric data derived from facial images is the unique mathematical characteristics that are extracted from the image in order to capture the individual identity. Those unique mathematical characteristics can then be compared to the characteristics extracted from other facial images to determine if there is a match. *See* Dr. Joseph. J. Atick, International Biometrics & Identification Association, *Face Recognition in the Era of Cloud and Social Media: Is it Time to Hit the Panic Button?* (Dec. 2011), at 2, *available at* http://www.ibia.org/resources.

8. *See Face Facts Workshop, Remarks of Dr. Jonathan Phillips, NIST*, at 23-24.

9. *See id.* These tests were done with a limited set of frontal images that were controlled for illumination; the same results could not necessarily be duplicated with snapshots taken on the street or photos posted on social networks, many of which do not contain ideal pose or lighting conditions. *See id.* at 29.

10. *See id.* at 32.

face to an identified "reference photo," where the name of the individual is known. Until recently, it was difficult to match two images if the photos were taken from different angles. With current technologies, companies can generate 3D face images to help reconcile pose variations in different images.[11]

These recent technological advances have been accompanied by rapid growth in the availability of photos online.[12] Panelists noted that ten years ago, most of the images available online were of celebrities, while today there are many sources of identified images of private citizens online.[13] One explanation for this is the rise in popularity of social networking sites. For example, in a single month in 2010, 2.5 billion photos were uploaded to Facebook.[14] This multitude of identified images online can eliminate the need to purchase proprietary sets of identified images, thereby lowering costs and making facial recognition technologies commercially viable for a broader spectrum of commercial entities.[15]

2. Current commercial uses of facial recognition technologies

Facial recognition technologies currently operate across a spectrum ranging from facial detection, which simply means detecting a face in an image, to individual identification, in which an image of an individual is matched with another image of the same individual. In the latter example, if the face in either of the two images is identified – *i.e.* the name of the individual pictured is known – then, in addition to demonstrating a match between two faces, the technology can be used to identify previously anonymous faces. In between these two divergent uses are a range of possibilities that include determining the demographic characteristics of a face, such as age range and gender, and recognizing emotions from facial expressions.[16]

---

11. *Face Facts Workshop, Remarks of Dr. Ralph Gross, Carnegie Mellon University*, at 20.

12. The increasing availability of identified images online is important because it allows facial recognition systems to not only match two images of the same individual, but identify that individual by name. *See also Comment of the Center for Democracy & Technology*, cmt. #87, 3.

13. *See Face Facts Workshop, Remarks of Prof. Alessandro Acquisti, Carnegie Mellon University*, at 133, 139-140.

14. *See id.* at 140.

15. *See Comment of the Center for Democracy & Technology*, cmt. #87, 3; *see also Face Facts Workshop, Remarks of Dr. Ralph Gross, Carnegie Mellon University*, at 33-34 (having multiple images of a subject allows the systems to overcome difficulties such as bad lighting or a bad pose that may affect particular images).

16. *See* Todd Bishop, *Happy or sad? You might not see that ad, if Microsoft Kinect can figure out your mood*, GeekWire, June 10, 2012, *available at* http://www.geekwire.com/2012/happy-sad-microsoft-system-target-ads-based-emotional-state; Karen Weintraub, *But How Do You Really Feel? Someday the Computer May Know*, N.Y. Times, Oct. 15, 2012, *available at* http://www.nytimes.com/2012/10/16/science/affective-programming-grows-in-effort-to-read-faces.html.

Current uses of facial detection include, among others, refining search engine results to include only those results that contain a face, locating faces in images in order to blur or de-identify them, or ensuring that the frame for a video chat feed actually includes a face.[17] Facial detection is also used in virtual eyeglass fitting systems and virtual makeover tools that allow consumers to "try on" a pair of glasses or a new hairstyle online. In these systems, after the consumer has uploaded a photo of herself to the website, that photo is scanned, basic facial features are picked out and – using the detected facial features as reference points – the eyewear or hairstyle is superimposed on the consumer's face.[18]

More sophisticated technologies that not only distinguish a face from surrounding objects, but also assess various characteristics of that face, can be used commercially in a variety of ways. For instance, technologies that identify moods or emotions from facial expressions can be used to determine a player's engagement with a video game or a viewer's excitement during a movie.[19] Further, technologies that can determine the gender and age range of the person standing in front of a camera can be placed into digital signs or kiosks, allowing advertisers to deliver an advertisement in real-time based on the demographic of the viewer.[20] This could provide substantial benefits to advertisers by allowing them to quickly show relevant products and deals, possibly leading to more sales.[21]

One company – called SceneTap – has also leveraged the ability to capture age range and gender to determine the demographics of the clientele of bars and nightclubs.[22] Both the

---

17. *See Face Facts Workshop, Remarks of Benjamin Petrosky, Google*, at 108-110 (an image search on Google's search engine can be refined to include only image results that contain a face; its Street View service uses facial detection to blur faces that are found in its images); *Face Facts Workshop, Remarks of Gil Hirsch, face.com*, at 120-121 (face.com uses facial detection to ensure that there is actually a face in the frame for video chat feeds as a way to prevent sexually explicit video chatting). At the time of the Face Facts workshop, face.com was an independent provider of facial recognition technologies for developers. In June 2012, face.com was acquired by Facebook. Ari Levy, *Facebook Buys Face.com, Adds Facial Recognition Software*, BLOOMBERG, June 18, 2012, *available at* http://www.bloomberg.com/news/2012-06-18/facebook-buys-face-com-adds-facial-recognition-software.html.

18. *See e.g.*, Ray Ban, Ray Ban Virtual Mirror, http://www.ray-ban.com/usa/science/virtual-mirror; InStyle, Hollywood Makeover, http://www.instyle.com/instyle/makeover/0,,,00.html.

19. *See Face Facts Workshop, Remarks of Jai Haissman, Affective Interfaces*, at 59-60.

20. *See Face Facts Workshop, Remarks of Brian Huseman, Intel*, at 41, 43 (Intel developed its AIM suite software that includes these technologies and has been working with large brands such as Kraft and Adidas to place it into their digital signage).

21. A representative of Adidas noted: "If a retailer can offer the right products quickly, people are more likely to buy something." Shan Li and David Sarno, *Advertisers start using facial recognition to tailor pitches*, LA TIMES, Aug. 21, 2011, *available at* http://articles.latimes.com/2011/aug/21/business/la-fi-facial-recognition-20110821.

22. *See Face Facts Workshop, Remarks of Andrew Cummins, SceneTap*, at 66-68 (SceneTap uses Intel's AIM Suite software in its cameras to gather this demographic information).

operators of the venue and third parties – such as liquor distributors – can use facial data to understand the demographics of a particular venue's customers at certain times, and possibly tailor their specials or promotions accordingly.[23] SceneTap also makes the aggregate information it collects available through a mobile app that consumers can use to make decisions about which venues to patronize.[24] While these implementations do more than simply detect a face in an image, they do not derive unique biometric data for comparison purposes.

Technologies that do derive unique biometric data for comparison and identification have been implemented in a variety of manners. For example, they can be used for authentication purposes by enabling a mobile phone user to use her face, rather than a password, to unlock her phone.[25] One of the most prevalent current uses of this technology is to enable semi-automated photo tagging or photo organization on social networks and in photo management applications.[26] As currently implemented, these features on social networks only suggest "tags" of people that the user already knows, either through a "friend" relationship or other contacts that suggest the two individuals know each other.[27]

3. Possible future uses of facial recognition technologies

In addition to discussing current uses of facial recognition technologies, workshop panelists discussed ways in which companies could implement these technologies in the future. Most of this discussion centered around the possibility that it may become feasible to use facial recognition to identify anonymous individuals in public places, such as streets or retail stores, or in unidentified photos online. While it does not seem that it is currently possible for commercial entities to accomplish this on a wide scale, recent studies suggest that in the near future, it may

---

23. *Id.* at 67-70.

24. *Id.*

25. *See Face Facts Workshop, Remarks of Benjamin Petrosky, Google*, at 110-111 (Google has implemented this technology in its Face Unlock feature for Android devices).

26. *See id.* at 111-112 (Google has enabled facial recognition technology in both its Picasa photo management software and its social network, Google+); *Face Facts Workshop, Remarks of Gil Hirsch, face.com*, at 122 (prior to its acquisition by Facebook, face.com also provided users with the ability to tag photos on social networks); *Face Facts Workshop, Remarks of Erin Egan, Facebook,* at 222 (Facebook used facial recognition for its "Tag Suggest" feature). At the time of the Face Facts workshop, Facebook's "Tag Suggest" feature was active on the Facebook website. In 2012, Facebook suspended the feature and reportedly plans to restore it in the future. *See* Statement of Rob Sherman, Manager of Privacy and Public Policy, Facebook, *What Facial Recognition Means for Privacy and Civil Liberties: Hearing before the S. Subcomm. On Privacy, Tech. and the Law,* 112th Cong. (July 18, 2012) *available at* http://www.judiciary.senate.gov/pdf/12-7-18ShermanTestimony. pdf. As of the date of this report, the feature has not been re-activated.

27. *See Face Facts Workshop, Remarks of Erin Egan, Facebook*, at 222; *Face Facts Workshop, Remarks of Gil Hirsch, face.com*, at 125; *Face Facts Workshop, Remarks of Benjamin Petrosky, Google*, at 153-156.

be possible. For example, in a 2011 study, Carnegie Mellon researchers were able to identify individuals in previously unidentified photos from a dating site, by using facial recognition technology to match them to their Facebook profile photos.[28]

4. Privacy concerns raised by current and possible future uses of facial recognition technologies

As illustrated by the above examples, companies can use facial recognition technology in ways that benefit consumers by providing them innovative products and services, such as the ability to try beauty products by uploading their faces to the Web, the ability to target search results, and the ability to organize and manage photos. Companies can also use the technology to protect privacy, by, for example, detecting and blurring images in photos, or using faces instead of passwords as an authentication device to unlock mobile phones.

At the same time, the use of facial recognition technologies can raise privacy concerns. For example, panelists voiced concerns that databases of photos or biometric data may be susceptible to breaches and hacking.[29] Further, panelists discussed how some consumers may perceive digital signs equipped with cameras using facial recognition technologies as invading their privacy because they can detect consumers from a distance and process their images without their knowledge or consent.[30] Unless these signs are labeled, they often look no different to consumers than digital signs that are not equipped with cameras. Panelists representing companies that currently use facial recognition technologies similarly acknowledged that there are privacy concerns surrounding the use of these technologies. For example, a Google representative noted the company's reluctance to implement facial recognition until it had put appropriate privacy protections in place.[31]

---

28. This study used a limited geographic area, and therefore a limited number of photos and subjects; thus, the results cannot necessarily be duplicated on larger scale. *See Face Facts Workshop, Remarks of Prof. Alessandro Acquisti, Carnegie Mellon University*, at 130-131, 138-139. The researchers also had some success in determining the identity of students walking on a college campus by using facial recognition to match the photo taken of them walking on campus with their publicly available photos on Facebook. Further, the researchers were able to use publicly available information about the individuals they identified to indicate the likely first five digits of their social security number. These experiments used cooperative subjects willing to stop and have their picture taken; it is likely not yet possible to recreate them in order to identify anyone on the street at any time. *Id.*

29. *See Face Facts Workshop, Remarks of Chris Conley, ACLU of Northern California*, at 166.

30. *See Face Facts Workshop, Remarks of Fred Carter, Information and Privacy Commissioner's Office, Ontario, Canada*, at 77; *Face Facts Workshop, Remarks of Beth Givens, Privacy Rights Clearinghouse*, at 85.

31. *See Face Facts Workshop, Remarks of Benjamin Petrosky, Google*, at 107.

Perhaps of most concern, panelists surmised that advances in facial recognition technologies may end the ability of individuals to remain anonymous in public places.[32] For example, a mobile app that could, in real-time, identify anonymous individuals on the street or in a bar could cause serious privacy and physical safety concerns, although such an app might have benefits for some consumers. Further, companies could match images collected by digital signs with other information to identify customers by name and target highly-personalized ads to them based on past purchases, or other personal information available about them online.[33] Social networks could identify non-users of the site – including children – to existing users, by comparing uploaded images against a database of identified photos. Although staff is not aware of companies currently using data in these ways, if they begin to do so, there would be significant privacy concerns.

B. Public Comments

Following the Face Facts workshop, Commission staff requested public comments regarding a number of topics and questions.[34] Among other issues, commenters were asked to provide input on: how consumers can benefit from facial recognition technologies; the privacy and security concerns surrounding the commercial use of these technologies; best practices for providing consumers with notice and choice about the use of these technologies; and best practices for deploying these technologies in a way that protects consumer privacy. The comments received reflected a wide variety of viewpoints on these issues, and are discussed below.[35]

---

32. *See Face Facts Workshop, Remarks of Chris Conley, ACLU of Northern California*, at 149-150. Professor Acquisti also noted the possibility that this could be accomplished by incorporating facial recognition technology into mobile phone cameras, or even glasses or contact lenses. *See Face Facts Workshop, Remarks of Prof. Alessandro Acquisti, Carnegie Mellon University*, at 142-143.

33. *See Comment of the Center for Democracy & Technology*, cmt. #87, 8 ("[I]t is likely that digital media will one day routinely identify individuals for the simple reason that it will be profitable to do so."); *see also* Stephanie Clifford, *Instant Ads Set the Pace on the Web*, N.Y. TIMES (March 12, 2010), *available at* http://www.nytimes.com/2010/03/12/business/media/12adco.html (noting that the online advertising industry has been willing to pay a premium for personalized advertisements).

34. *See* Press Release, FTC, FTC Seeks Public Comments on Facial Recognition Technology (Dec. 23, 2011), *available at* http://www.ftc.gov/opa/2011/12/facefacts.shtm.

35. *See* FTC, FTC Seeks Public Comments on Facial Recognition Technology; Project Number P115406, http://www.ftc.gov/os/comments/facialrecognitiontechnology/index.shtm.

III. General Themes

Panelists and commenters discussed several general themes that apply across the spectrum of facial recognition technologies. First, many panelists and commenters agreed that companies should implement privacy protections for all facial recognition technologies, including those that do not individually identify consumers. Commenters and panelists noted that, despite the fact that a person's face is public in the sense that numerous strangers may see it on a daily basis, it is also a persistent, unique identifier that consumers do not have the ability to replace.[36] As such, it is inextricably linked to a consumer and deserving of privacy protections.[37] FTC staff agrees with commenters who stated that companies using facial recognition technologies that operate anywhere along the spectrum – from detection to deriving unique biometric data for comparison purposes – should implement privacy protections appropriate for the context of their relationship with consumers. This conclusion is consistent with the Commission's Privacy Report, which stated that commercial entities should implement privacy protections for any data that can be reasonably linked to a specific consumer, computer, or other device.[38]

Second, a significant number of panelists and commenters agreed that it is important for companies using these technologies to increase the transparency of their data practices, given that the commercial use of the technologies is relatively new and can often be invisible to consumers.[39] To this end, panelists and commenters noted that increased consumer education about the use of facial recognition technologies is of paramount importance and that all stakeholders – including industry, trade associations, consumer and privacy groups, and government entities – should engage in consumer education efforts. FTC staff agrees that consumer education is very important. In addition to the uses and implications of facial recognition technologies, these expanded education efforts should include topics such as the

---

36. *See Face Facts Workshop, Remarks of Prof. Alessandro Acquisti, Carnegie Mellon University*, at 135 ("Your face is a veritable conduit between your different online personas; in fact, between your offline and online persona. It's much easier to change your name and declare 'reputational bankruptcy' than to change your face. Your face creates the link between all the different personas."); *Comment of The Electronic Privacy Information Center*, cmt. #83, 12 ("unlike a credit card or social security number, it's not possible to go out and get a new faceprint if your biometric data is hacked").

37. *See Comment of Software & Information Industry Association*, cmt. #77, 4 ("if face print information is retained, this creates a reasonable possibility of identification").

38. There is a carve-out for entities that collect only non-sensitive data from fewer than 5,000 consumers per year and that do not share the data with third parties.

39. *See, e.g., Face Facts Workshop, Remarks of Chris Conley, ACLU of Northern California*, at 147-149; *Face Facts Workshop, Remarks of Beth Givens, Privacy Rights Clearinghouse*, at 85; *Face Facts Workshop, Remarks of Fred Carter, Information and Privacy Commissioner's Office, Ontario, Canada*, at 77.

potential implications of making photos publicly available.[40] This topic has particular relevance for teens, who often impulsively post photos and other content without an understanding that the photos could be used for unintended, secondary purposes.

Third, several panelists and commenters discussed the role of self-regulation in this area. Participants noted that trade groups representing companies using facial recognition technologies, most notably in the digital signage industry, have proactively issued guidance and "best practices" for their members. For example, Point of Purchase Advertising International's Digital Signage Group ("POPAI") has developed a code of conduct containing recommendations for marketers to follow in order to maintain ethical data collection practices in retail settings.[41] Similarly, the Digital Signage Federation worked with the Center for Democracy and Technology to craft a voluntary set of privacy guidelines for their members, which include advertisers and digital sign operators.[42] Both of these self-regulatory codes address the use of facial recognition technologies in digital signs. FTC staff supports these efforts and encourages trade associations to explore ways to enforce compliance with their voluntary codes of conduct and privacy standards. The Commission will also continue to enforce the FTC Act against companies that engage in unfair or deceptive practices, including by failing to abide by self-regulatory programs they join.

Finally, panelists and commenters discussed the need for companies implementing facial recognition to consider and implement privacy protections. In many cases, the protections suggested by panelists are covered by the principles articulated in the March 2012 Privacy Report, which are privacy by design, simplified choice, and improved transparency. To this end, the Commission staff has developed a series of case studies describing how companies can implement these principles in specific facial recognition scenarios, as described below.

---

40. *See Face Facts Workshop, Remarks of Erin Egan, Facebook*, at 204.

41. *See* Press Release, POPAI, Taking Consumer Privacy Seriously, POPAI's Digital Signage Group Releases Code of Conduct (Aug. 2, 2010), *available at* http://www.popai.com/2010/02/08/taking-consumer-privacy-seriously-popais-digital-signage-group-releases-code-of-conduct/

42. Digital Signage Federation, *Digital Signage Privacy Standards* (Feb. 2011), *available at* http://www.digitalsignagefederation.org/Resources/Documents/Articles%20and%20Whitepapers/DSF%20Digital%20Signage%20Privacy%20Standards%2002-2011%20%283%29.pdf. *Face Facts Workshop, Remarks of Harley Geiger, the Center for Democracy and Technology*, at 52.

IV. Case Studies on Common Commercial Uses of Facial Recognition Technologies

This report contains three case studies, each focusing on a common commercial use of facial recognition technologies. The case studies are not intended to represent an exhaustive discussion of every current use of these technologies, nor every possible use that may occur in the future. They are merely examples that demonstrate suggested best practices for common uses and represent a spectrum of increasing sophistication of the technology involved. They build on the discussions at the workshop, comments received, and principles contained in the March 2012 Privacy Report.

Case Study #1: Facial Detection

> *Scenario:* An eyeglass company allows consumers to upload their images to the company's website and then uses facial detection to detect the face and eyes in the image and superimpose various styles of glasses on the consumer's face. The company stores the images in order to enable consumers to use this feature of the website in the future without uploading a new image.

In this scenario, the eyeglass company's use of facial detection to simply locate the consumer's face in an image the consumer has voluntarily uploaded does not, by itself, raise privacy concerns. However, there may be privacy concerns related to the company's collection and storage of the consumer's image in connection with its virtual fitting feature. The eyeglass company should take several steps to address these concerns.

First, it should design its service with privacy in mind by implementing "privacy by design." For instance, Commission staff agrees with panelists who noted that companies that hold databases of consumer images should take steps to protect those images.[43] In this example, the eyeglass company should have reasonable data security protections in place for any stored images in order to prevent unauthorized access to those images. In addition, the eyeglass company should implement a specified retention period and dispose of stored images once they are no longer necessary for the purpose for which they were collected. If a consumer deletes his

---

43. *See Face Facts Workshop, Remarks of Erin Egan, Facebook*, at 229; *see also* Dr. Joseph J. Atick, International Biometrics & Identification Association, *Face Recognition in the Era of the Cloud and Social Media: Is it Time to Hit the Panic Button* (Dec. 2011), at 5-6, *available at* http://ibia.org/resources/whitepapers/.

or her account on the website, the stored images are no longer necessary and should be disposed of, even if the stated retention period has not yet passed.

Second, consistent with panelist and commenter recommendations on transparency and choice, the company should be clear with consumers about its data practices and provide choices appropriate for the context of the transaction. One appropriate way to achieve transparency in this scenario is to state – at the time the consumer uploads her image to try on glasses – why the company is storing, rather than immediately deleting, that image (*e.g.*, "Create an account so you can try on glasses in the future"). If the company is storing the images for a purpose that is not consistent with the context of the transaction taking place, it should provide additional information about why it is storing the images – at a "just in time" point. For example, if the company stores the images for purposes of sharing them with third parties, it should explicitly provide consumers with a choice about this practice before they upload their image – outside of a privacy policy or similar document. In all cases, the company should also inform consumers of: (1) the length of time the images are stored, (2) who will have access to the stored images, and (3) consumers' rights regarding deletion of the stored images.

Finally, consistent with the recommendations in the Commission's Privacy Report, if in the future the eyeglass company decides to use the images in a materially different manner than it represented at the time of collection, the company should obtain the affirmative express consent of the consumer prior to such use. For instance, if the eyeglass company decides to use the images in its advertising, rather than simply storing them for future use by the consumer, this would require the consumer's affirmative express consent.

Case Study #2: Detection or Recognition of Demographic Characteristics in Digital Signs

Scenario: A sports drink company operates digital signs in a supermarket. These signs include cameras and have the ability to assess the age range and gender of the consumer standing in front of them. The signs display a targeted advertisement to the consumer based on those demographic characteristics.[44] The consumer's image is processed instantaneously while the consumer is standing in front of the sign and is not stored for future use.

In this scenario, the sports drink company should implement privacy by design in several ways. First, even though it is not storing images, it should still take steps to implement reasonable data security protections to protect against the possibility that a third party could hack into the sign's software and access the images in real-time. During the workshop, existing companies using facial recognition technologies for demographic detection discussed employing these types of protections. For example, SceneTap is a company that sets up cameras in bars to determine the aggregate age range and gender of a venue's patrons. It secures the feed from its cameras so that the feed cannot be accessed by anyone in the venue.[45] Staff believes that all companies operating digital signs should follow this type of approach.

Second, the sports drink company in this scenario should consider carefully where to place such signs and avoid placing them in sensitive areas. This recommendation is consistent with recommendations by both consumer groups and industry trade associations. For example, the World Privacy Forum, along with other consumer advocacy groups, drafted the Digital Signage Privacy Principles, which contain recommended consumer protection measures for digital signage using facial recognition technologies.[46] These principles state that such digital signage should not be placed in bathrooms, locker rooms, health care facilities, or areas where children congregate.[47] Similarly, POPAI's code of conduct recommends that companies not use "observed

---

44. Not all digital signs contain cameras or are equipped with facial recognition technologies. The recommendations contained within this case study are intended to apply only to digital signs that use facial recognition technologies on individuals standing in front of a camera or sensor placed in the sign.

45. *See Face Facts Workshop, Remarks of Andrew Cummins, SceneTap,* at 71.

46. *See Comment of World Privacy Forum,* cmt. #82, 4-5.

47. *See id.* at 5.

tracking data" in HIPAA-compliant areas, such as pharmacies.[48] Commission staff agrees that digital signage companies should observe these limitations, and be vigilant regarding the locations of such signs.

Third, in this scenario, the company is not storing the images. Staff considers this a best practice for companies using digital signs and other applications using similar demographic detection. As one panelist noted, because the consumer's facial characteristics are analyzed and an advertisement delivered in a matter of seconds, the underlying image quickly loses its usefulness.[49] Much of the additional data that can be derived from these digital signs, such as which demographic groups spent more time looking at particular advertisements, or how many consumers walked by a particular sign, can be stored in the form of aggregate statistics as opposed to images, thereby eliminating or at least greatly reducing the risk that the data will be identified or tied to a particular individual.[50]

With respect to choice and transparency, panelists at the Face Facts Workshop noted that consumers likely do not currently expect signs to detect their age range and gender and target an advertisement to them in real-time on the basis of these attributes.[51] Indeed, as some commenters and panelists noted, providing a clear notice is particularly important because a digital sign or kiosk that contains a camera using facial recognition technologies will often look no different to a consumer than a digital sign that does not have a camera within the display.[52] Choice is important in these situations as well. Staff agrees with commenters and panelists who noted

---

48. *See* POPAI, Digital Signage Group, *Best Practices: Recommended Code of Conduct for Consumer Tracking Research* (Feb. 2010), at 6.

49. *See Face Facts Workshop, Remarks of Simon Rice, Technology Information Commissioner's Office, United Kingdom*, at 220-21.

50. *See Face Facts Workshop, Remarks of Brian Huseman, Intel*, at 41, 44 (discussing the benefits aggregate statistics can provide to advertisers).

51. *See Face Facts Workshop, Remarks of Fred Carter, Information and Privacy Commissioner's Office, Ontario, Canada*, at 77 ("you are detected whether you know it or not. And this has very important implications for applying fair information practices. You don't know it's happening. It's taken at a distance. You have no knowledge."); *Face Facts Workshop, Remarks of Harley Geiger, The Center for Democracy and Technology*, at 56 ("I have seen in press reports companies literally saying, literally declining to point out which signs actually have facial detection or facial recognition and 'say they don't want their customers to feel uncomfortable.'").

52. *See Face Facts Workshop, Remarks of Fred Carter, Information and Privacy Commissioner's Office, Ontario, Canada*, at 77; *see also, Comment of the Center for Democracy and Technology*, cmt. #87, 5 (noting that many digital signs using this technology are not labeled). However, at least one commenter suggested that notice may not be required when digital signs do not uniquely identify individuals. *See Comment of TechAmerica*, cmt. #79, 3. *But see Comment of the Center for Democracy and Technology*, cmt. #87, 9 (noting that transparency provides benefits to both consumers and businesses alike, because "[s]ecrecy magnifies consumers' sense that their privacy is being invaded – if companies try to hide the fact that they are using facial recognition, it will sensationalize the issue and lead consumers to more deeply distrust the technology").

that because the use of these technologies within digital signs is not currently consistent with reasonable consumer expectations, consumers should be offered a choice as to whether they wish to come into contact with a digital sign that markets to them based upon their perceived age range and gender.

Applying these principles to the example of the sports drink company, the company should provide clear notice that digital signs using facial recognition to detect demographic characteristics are in operation, before the consumer comes into contact with the sign. This way the consumer can choose to avoid the sign. Depending upon the size of the store, the notice may be a prominent notice at the entrance to the store itself or at the entrance to a particular section of the store – and on or near the sign itself.[53] Further, as panelists and commenters stated, a bare listing of a website or company name or logo on the notice alone would not be sufficient to inform consumers.[54] At a minimum, a notice should clearly state the purpose of the technology and indicate how consumers can find more information about the technology and the practices of the company operating the signs in that venue.[55] A consumer who does not wish to have their data used in this manner is then able to choose not to shop at this particular store or avoid the location where the sign is placed.

Panelists and commenters discussed the appropriateness of this type of "walk away" choice. Some panelists stated that it is unfair for consumers to have to avoid a retail location altogether in order to avoid these types of digital signs.[56] Other panelists suggested that the

---

53. *See Face Facts Workshop, Remarks of Harley Geiger, Center for Democracy and Technology,* at 53-54 (suggesting layered privacy notices, beginning with a comprehensive privacy policy on the website of the owner of the device, as well as the website of the location where the camera is located; an abbreviated notice at the perimeter of the area where the technology is in use; and a short notice on the actual sign to alert consumers that the sign is using facial recognition technologies).

54. *See Face Facts Workshop, Remarks of Pam Dixon,* at 179; *see also, Comment of the Center for Democracy and Technology,* cmt. #87, 16 ("While a symbol may one day alert consumers to the presence of a facial recognition or detection device or program, a symbol will only be an adequate form of notice if it is adopted on an industry-wide basis and consumers are properly educated on the meaning of the symbol").

55. One panelist suggested that the notices in place at the venue or on the device could include a QR code that would allow consumers who wanted more in depth information to use their smart phone to obtain it. *See Face Facts Workshop, Remarks of Beth Givens, Privacy Rights Clearinghouse,* at 90-91.

56. *See, e.g., Face Facts Workshop, Remarks of Pam Dixon, World Privacy Forum,* at 183 (stating that "we shouldn't have to live in an opt-out village"); *Face Facts Workshop, Remarks of Beth Givens, Privacy Rights Clearinghouse,* at 85-86 (stating that it is unfair for consumers to miss out on opportunities because they are uncomfortable with digital signage).

privacy risks associated with digital signs that do not store images or individually identify consumers are relatively low, and therefore this level of choice is appropriate.[57]

FTC staff supports a sliding scale approach to notice and choice. Provided that companies are sufficiently transparent, only detect age and gender, and do not store consumers' images, staff believes that a "walk away choice" may be sufficient. However, staff also encourages companies to consider other options, such as employing these technologies in kiosks that require consumer interaction prior to processing the consumer's image. For instance, Kraft Foods is reportedly considering installing face scanning kiosks in supermarkets.[58] Alternatively, companies could place digital signs only in limited sections of their stores. Staff believes that if companies begin storing consumers' images, or tracking consumers across signs, the privacy risks become much greater, and therefore the companies should provide consumers with more robust transparency and choices.

Finally, as several panelists suggested, if the sports drink company decides to begin individually identifying consumers through digital signs – such as by running those images against a database of images identified by name – the company should first obtain the consumer's affirmative express consent.[59] Obtaining the consumer's affirmative express consent is the only way to give the consumer a *meaningful* choice in this scenario, where a company is using facial recognition to identify a consumer who it could not otherwise identify. As one commenter noted, without facial recognition technology, "most individuals in public may expect that few businesses and passersby would recognize the individual's face, fewer would affix a name to the face, and fewer still would be able to associate the face with internet behavior… or other profiles."[60] The commenter further noted that, "facial recognition technology can fundamentally change that dynamic, enabling any marketer… to collect – openly or in secret – and share the identities and associated personal information of any individual whose face is captured by the camera."[61] Commission staff agrees with these observations, and it is staff's position that, before

---

57. *Face Facts Workshop, Remarks of Harley Geiger, Center for Democracy and Technology*, at 54 (stating that for these types of signs consumers can exercise choice by choosing not to shop in particular venues upon being notified of the practices).

58. Shan Li and David Sarno, *Advertisers start using facial recognition to tailor pitches*, LA TIMES, Aug. 21, 2011, *available at* http://articles.latimes.com/2011/aug/21/business/la-fi-facial-recognition-20110821.

59. *See, e.g., Face Facts Workshop, Remarks of Harley Geiger, Center for Democracy and Technology*, at 54-55; *Face Facts Workshop, Remarks of Dr. Joseph J. Atick, International Biometrics & Identification Association*, at 205.

60. *Comment of the Center for Democracy and Technology*, cmt. #87, 7.

61. *Id.*

using facial recognition to identify an individual it could not otherwise identify, the company should obtain the affirmative express consent of the individual in the image.

Case Study #3: Facial Recognition in Online Social Networks

Scenario: An existing social network implements a facial recognition feature. When a user uploads new photos, the social network scans those photos against existing "tagged" photos of the user's "friends."[62] The social network then identifies the user's "friends" in the new photos so the user can tag them. Users cannot utilize the feature to identify other users who are not their "friends."

The social network in this scenario should engage in privacy by design by implementing several privacy protections. First is data security, both for the images themselves as well as the biometric data derived from the images.[63] Several panelists and commenters focused on the importance of proper data security practices for companies storing biometric data.[64] One protection that has been implemented by some companies and recommended by commenters is encryption of the stored data.[65] Facebook, for example, encrypted the data that it derived from images and used to make comparisons and suggest tags.[66] Commission staff agrees that companies using facial recognition technologies should appropriately protect the images and biometric data they store.

Moreover, even if a company does not itself intend to implement facial recognition technologies, it should consider putting protections in place that would prevent unauthorized scraping of the publicly available images it stores in its online database. As discussed above, the increasing availability of identified images online has been a major factor in the increasing

---

62. Throughout this scenario, staff uses the term "friend" to refer to an individual user that another user has a direct mutual connection with on the social network.

63. *See United States of America (For the Federal Trade Commission) v. RockYou, Inc.*, Case No. 3:12-cv-01487-SI (Mar. 27, 2012) (consent decree), *available at* http://www.ftc.gov/os/caselist/1023120/index.shtm (the FTC alleged that RockYou failed to live up to its representation that it would take reasonable and appropriate measures to safeguard its customers personal information – including photographs – from unauthorized access).

64. For example, one commenter raised the concern that a database containing biometric information derived from facial images could become a target for hackers and possibly lead to identity theft. *See Comment of Electronic Privacy Information Center*, cmt. #83, 12; *see also, e.g., Face Facts Workshop, Remarks of Chris Conley, ACLU of Northern California*, at 166; *Face Facts Workshop, Remarks of Dr. Joseph J. Atick, International Biometrics & Identification Association*, at 206; *Face Facts Workshop, Remarks of Erin Egan, Facebook*, at 223.

65. *See Comment of Electronic Privacy Information Center*, cmt. #83, 21.

66. *See Comment of Facebook*, cmt. #81, 7.

commercial viability of facial recognition technologies.[67] Therefore, as commenters suggested, in order to prevent an unintended secondary use of those images, companies should put in place appropriate precautions.[68]

Second, the social network described in this scenario should establish and maintain appropriate retention and disposal practices. Panelists and commenters noted that several social networking companies have implemented processes to delete biometric data when it is no longer necessary. For example, Google+'s "Find My Face" feature uses facial recognition technology to suggest tags of other users that have turned on the feature and that the user uploading the photo is connected with in some way.[69] If a user decides to turn the feature off after previously opting in, the biometric data collected about that user is deleted.[70]

In this scenario, the social network should also be transparent with consumers about its data practices regarding the facial recognition feature and provide consumers a choice about the use of facial recognition technologies.[71] Commenters and panelists noted that consumers generally provide photos and images to social networks for the purpose of organizing them and sharing them with friends.[72] While providing innovative ways to implement photo tagging may make the process of organizing and sharing photos with friends more convenient, consumers may not currently expect that their biometric data would be collected from those images and stored by the social network. Because this use is not currently within the context of consumers' relationship with the social network, when the company first rolls out this feature and begins analyzing users' photos to gather biometric data, it should provide users with a clear notice, outside of a

---

67. *See supra* page 4.

68. *See* Dr. Joseph J. Atick, International Biometrics & Identification Association, *Face Detection & Face Recognition Consumer Applications: Recommendations for Responsible Use* (Dec. 2011), at 3, *available at* http://www.ibia.org/resources/ ("Protecting against building identity databases by harvesting the web requires the implementation of technical measures by those who control the repositories of social media images and search engine companies. For example, such harvesting can be prevented if the image servers block all web-crawlers that do not originate from search engine entities that have previously agreed to a declared privacy policy. Such a policy would include a commitment not to extract faceprints from these images or not to make them available for general search."); s*ee also, Face Facts Workshop, Remarks of Erin Egan, Facebook*, at 204 (noting that Facebook's Statement of Rights and Responsibilities prohibits unauthorized scraping).

69. *See Face Facts Workshop, Remarks of Benjamin Petrosky, Google,* at 112-113.

70. *See id.* at 114. Similarly, if a user opted out of Facebook's "Tag Suggest" feature, Facebook deleted any previously collected biometric data. *See Comment of Facebook,* cmt. #81, 6.

71. *See Comment of Software & Information Industry Association*, cmt. #77, 5 (stating that the use for photo tagging on social networks should be under the control of the data subject who should be allowed to opt-in or opt-out).

72. *See Comment of Facebook*, cmt. #81, 2; *Face Facts Workshop, Remarks of Benjamin Petrosky, Google,* at 112-114.

privacy policy, about how the feature works, what data it collects, and how that data will be used. Companies should also provide consumers with an easy to find, meaningful choice not to have their biometric data collected and used for facial recognition. Further, consumers should be able to turn off the feature at any time and delete any biometric data previously collected from their tagged photos.

Finally, the social network should not collect and store biometric data of non-users of its service because there is no context in which to provide such non-users with a choice about these practices. Whereas the social network can give users of its service an opportunity to control the use of facial recognition technology, there is no practical way for the social network to offer such choice to non-users.[73]

There are at least two scenarios in which social networks should obtain consumers' affirmative express consent before collecting or using biometric data from facial images. First, as with all companies, social networks should obtain a consumer's affirmative express consent before using a consumer's image or any biometric data derived from that image in a materially different manner than it represented when it collected the data. Second, the social network should not identify users to other users who are not their "friends" on the site without first obtaining their affirmative express consent. Identifying an anonymous or unidentified individual in an image to someone who did not already know their identity is outside the context of the individual's relationship or transaction with the social network. In fact, at the workshop a representative of Facebook pointed out the importance of the context in which their facial recognition feature operated, stating, "we have basic principles around our use of it. Number one, it's within the social context. So, we are not using facial recognition technology to identify people who are not known to you."[74] Offering choice on an opt-out basis would be ineffective for this use because once a person has been identified to a stranger, he or she cannot be un-identified after the fact. A consumer's face is a persistent identifier that cannot be changed in the way that a consumer could get a new credit card number or delete a tracking cookie. Consequently, in order to offer meaningful choice to consumers about these practices, social networks should obtain

---

73. *See Face Facts Workshop, Remarks of Gil Hirsch, face.com,* at 158-160. Further, as panelists at the workshop noted, an opt-out that requires consumers to provide photos of themselves so that a service will know not to collect their biometric data or recognize them in any images going forward would itself raise privacy concerns. *See id.,* at 158-159; *Face Facts Workshop, Remarks of John Verdi, Electronic Privacy Information Center,* at 219 (while discussing the feasibility of a "do not track" option for facial recognition, Mr. Verdi stated, "you need to be tracked in order to assert your right not to be tracked. There needs to be a generation of [biometric data] in order to compare it against the do not track database.").

74. *See Face Facts Workshop, Remarks of Erin Egan, Facebook,* at 221.

consumers' affirmative express consent before identifying their anonymous images to users who could not otherwise identify them.

This last point has critical implications for the broader use of facial recognition technology. Even beyond social networks, companies should not use facial recognition to identify anonymous images of a consumer to someone who could not otherwise identify him or her, without obtaining the consumer's affirmative express consent. Consider the example of a mobile app that allows users to identify strangers in public places, such as on the street or in a bar. If such an app were to exist, a stranger could surreptitiously use the camera on his mobile phone to take a photo of an individual who is walking to work or meeting a friend for a drink and learn that individual's identity – and possibly more information, such as her address – without the individual even being aware that her photo was taken. Given the significant privacy and safety risks that such an app would raise, only consumers who have affirmatively chosen to participate in such a system should be identified.

V. Conclusion

The recent technological advances in the field of facial recognition undoubtedly provide a variety of benefits to consumers in the form of interesting products and services. However, as we have seen with other technologies, technological advances and the attendant business models they create often move faster than consumers' awareness or comfort. The best practices recommended in this report can guide companies as they develop new products and services and craft the processes and systems that will govern their operations. Moreover, because implementing these practices will promote consumer trust and ensure the continued growth of this industry, companies currently have incentives to engage in them. Fortunately, the commercial use of facial recognition technologies is still young. This creates a unique opportunity to ensure that as this industry grows, it does so in a way that respects the privacy interests of consumers while preserving the beneficial uses the technology has to offer. The FTC will continue to monitor this area and explore ways to work with industry, and educate and protect consumers.

Dissenting Statement of Commissioner J. Thomas Rosch

Facing Facts: Best Practices for Common Uses of Facial Recognition Technologies
October 22, 2012

I respectfully dissent from the issuance of the Staff Report entitled, "Facing Facts: Best Practices for Common Uses of Facial Recognition Technologies" ("Report" or "Staff Report"). Although I appreciate Staff's efforts to examine the issues surrounding the development and use of facial recognition technology, I believe the Report goes too far, too soon. My reasoning is threefold.

First, I object to the recommendations made in the Staff Report to the extent that they are rooted in Staff's insistence that the "unfairness" prong, rather than the "deception" prong, of the consumer protection portion of Section 5 of the Federal Trade Commission Act, should govern practices relating to facial recognition technology. Section 5(n) limits our unfairness authority to an act or practice that "causes or is likely to cause substantial injury to consumers which is not reasonably avoidable by consumers themselves and not outweighed by countervailing benefits to consumers or to competition."[1] As I have pointed out before, the Commission represented in its 1980 and 1982 Statements to Congress that it will generally enforce the consumer protection "unfairness" prong of Section 5 only where there is alleged tangible injury, not simply "[e]motional impact and other more subjective types of harm."[2] The Staff Report on Facial Recognition Technology does not – at least to my satisfaction – provide a description of such "substantial injury." Although the Commission's Policy Statement on Unfairness states

---

1. Federal Trade Commission Act Amendments of 1994, Pub. L. No. 103-312.

2. *See* Letter from the FTC to Hon. Wendell Ford and Hon. John Danforth, Committee on Commerce, Science and Transportation, United States Senate, Commission Statement of Policy on the Scope of Consumer Unfairness Jurisdiction (Dec. 17, 1980), *reprinted in International Harvester Co.*, 104 F.T.C. 949, 1073 (1984) ("FTC Policy Statement on Unfairness") *available at* http://www.ftc.gov/bcp/policystmt/ad-unfair.htm; Letter from the FTC to Hon. Bob Packwood and Hon. Bob Kasten, Committee on Commerce, Science and Transportation, United States Senate, *reprinted in* FTC Antitrust & Trade Reg. Rep. (BNA) 1055, at 568-570 ("Packwood-Kasten letter"); and 15 U.S.C. § 45(n), which codified the FTC's modern approach. Commission letter to Senators Packwood and Kastes reaffirming Statement (Mar. 5, 1982).

that "safety risks" may support a finding of unfairness,[3] there is nothing in the Staff Report that

indicates that facial recognition technology is so advanced as to cause safety risks that amount

to tangible injury. To the extent that Staff identifies misuses of facial recognition technology, the

consumer protection "deception" prong of Section 5 – which embraces both misrepresentations

and deceptive omissions – will be a more than adequate basis upon which to bring law

enforcement actions.

Second, along similar lines, I disagree with the adoption of "best practices" on the ground

that facial recognition *may* be misused. There is nothing to establish that this misconduct has

occurred or even that it is likely to occur in the near future. It is at least premature for anyone,

much less the Commission, to suggest to businesses that they should adopt as "best practices"

safeguards that may be costly and inefficient against misconduct that may never occur.

Third, I disagree with the notion that companies should be required to "provide

consumers with choices" whenever facial recognition is used and is "not consistent with the

context of a transaction or a consumer's relationship with a business."[4] As I noted when the

Commission used the same ill-defined language in its March 2012 Privacy Report, that would

import an "opt-in" requirement in a broad swath of contexts.[5] In addition, as I have also pointed

out before, it is difficult, if not impossible, to reliably determine "consumers' expectations" in

any particular circumstance.[6]

In summary, I do not believe that such far-reaching conclusions and recommendations

can be justified at this time. There is no support at all in the Staff Report for them, much less the

kind of rigorous cost-benefit analysis that should be conducted before the Commission embraces

such recommendations. Nor can they be justified on the ground that technological change will

occur so rapidly with respect to facial recognition technology that the Commission cannot

---

3.   *See supra* n.2, FTC Policy Statement on Unfairness at 3; *International Harvester Co.*, 104 F.T.C. at 1073.

4.   Report at 2.

5.   *See* Dissenting Statement of Commissioner J. Thomas Rosch, Issuance of Federal Trade Commission Report, *Protecting Consumer Privacy in an Era of Rapid Change: Recommendations for Businesses and Policymakers* (March 26, 2012), *available at* http://ftc.gov/speeches/rosch/120326privacyreport.pdf.

6.   *Id.*

adequately keep up with it when, and if, a consumer's data security is compromised or facial recognition technology is used to build a consumer profile. On the contrary, the Commission has shown that it can and will act promptly to protect consumers when that occurs.